ARCHES NATIONAL PARK
Geology Tour

Deborah Ragland, Ph.D.   images by Dennis Tasa

Published by Tasa Graphic Arts, Inc.

Arches National Park Geology Tour
ISBN 978-1-58256-054-0

ENHANCED AUDIO CD (CD-ROM PARTITION) LICENSE GRANT
You hereby accept a nonexclusive, nontransferable, permanent license to install and use the program ON A SINGLE COMPUTER at any given time. You may copy the program solely for backup or archival purposes in support of your use of the program on the single computer. You may not modify, translate, disassemble, decompile, or reverse engineer the program, in whole or part. This CD-ROM is intended for use on stand alone computers. Network installation is not supported.

WARRANTY LIMITS
The media is distributed on an "AS IS" basis, without warranty. Neither the author, copyright owner, or publisher shall have any liability to you or any other person or entity with respect to any liability, loss, or damage.

© 2010 by Tasa Graphic Arts, Inc.
1210B Salazar Road
Taos, NM 87571 U.S.A.
www.tasagraphicarts.com

Cover photo by Dennis Tasa. Delicate Arch, Arches National Park, Utah, U.S.A.

Book and Enhanced Audio CD published by Tasa Graphic Arts, Inc.
Images by Dennis Tasa, programming and production by Dan Pilkenton, Karen Tasa, and Cindy Robison.

All rights reserved. No part of this book or accompanying Enhanced Audio CD may be reproduced, in any form or by any means, without permission in writing from the publisher.

Printed in the United States of America
10   9   8   7   6   5   4   3   2   1

CONTENTS

IT'S ALIVE! Cryptobiotic Crust ... iv

GEOLOGY ... 1

Colorado Plateau ... 2
Geologic Map ... 4
Stratigraphic Column ... 5
Geology of the Most Important Formations and Members with Respect to Arch Formation ... 6
Color of the Rocks ... 8
What is an Arch and What is a Natural Bridge? ... 9
Stages in the Formation of the Arches ... 11

TOUR ... 17

VISITOR CENTER
Moab Fault ... 18

COURTHOUSE TOWERS
Queen Nefertiti ... 20
Park Avenue ... 21
La Sal Mountains Viewpoint ... 22
Courthouse Towers Viewpoint ... 23
Three Gossips ... 24
Sheep Rock ... 25

THE GREAT WALL
Petrified Dunes Viewpoint ... 26

THE WINDOWS SECTION
Balanced Rock ... 28
Garden of Eden ... 30
Parade of Elephants ... 32
Double Arch ... 34
North and South Windows ... 36
Turret Arch ... 38

WOLFE RANCH
Delicate Arch ... 40
Delicate Arch Viewpoint ... 42

FIERY FURNACE
Fiery Furnace Viewpoint ... 44
Sand Dune Arch ... 46
Fins near Sand Dune Arch ... 47

DEVILS GARDEN
Skyline Arch ... 48
Aerial Fins in Devils Garden ... 49
Tunnel Arch ... 50
Pine Tree Arch ... 51
Landscape Arch ... 52
Wall Arch ... 54
Partition Arch ... 56
Double O Arch ... 56

Glossary ... 58
References ... 60

IT'S ALIVE!
Cryptobiotic Crust *Audio track 01*

Cryptobiotic crusts are relatively rigid, but thin crusts of cyanobacteria, green and brown algae, mosses, and lichens with minor amounts of liverworts, fungi, and bacteria that are found in arid climates. These crusts are especially important on the Colorado Plateau as they provide stabilization of the soil. Cryptobiotic crusts can compose as much as 70 percent of the living matter on the surface of the soil.

The crusts in high deserts such as the one here at Arches help prevent erosion of the soils, help with moisture retention, and can fix nitrogen for other plants to use. In addition, the irregular surface can hold seeds from plants and give the seeds a chance to germinate. Cryptobiotic crusts compose entire ecosystems.

The cryptobiotic crusts are extremely fragile. Even a footprint can disrupt the micro-ecosystems so much that it may take as much as 100 years to recover. In the interest of preserving the fragile desert, the Park Service and all who work in deserts ask you to stay on walkways so that the crusts are not destroyed.

GEOLOGY

Colorado Plateau Audio track 02

Arches National Park is one of the many national, state, and local parks located on the Colorado Plateau, a unique physiographic province encompassing parts of four states—the Four Corners states of Colorado, New Mexico, Utah, and Arizona. Arches is one of five national parks located in Utah that is within the Colorado Plateau. Although the Plateau can be differentiated into six sub-provinces, all are characterized by arid to semi-arid climates and flat-lying rocks. Sandstones, limestones, and shales are the most common rock types on the Plateau and structures formed in these rocks have generally resulted from weathering. Weathering is defined as all of the physical and chemical changes that happen to rock formations when they are exposed to surface or near-surface conditions.

Arches National Park also owes its unique structures to another important geologic structure: the Paradox Basin. Approximately 310 to 305 million years ago, shallow seas that covered some parts of the North American continent were cut off from ocean circulation and became salty. As the water evaporated in the hot, dry climate of the Middle to Late Pennsylvanian and mountains of the Uncompahgre Uplift to the east eroded, salts and sediments accumulated in great thicknesses in the basin.

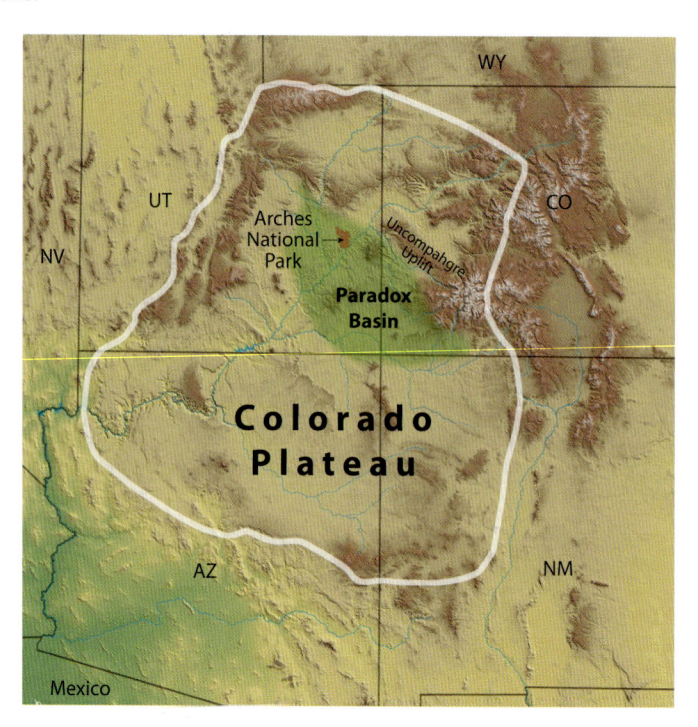

2

The thick salt layers were buried by other sediment and, because salt can behave in a very plastic manner, subsurface salt domes were formed. This contributed to some of the folding of rock layers above the salt domes.

During the late Cretaceous—about 70 to 80 million years ago—and extending into the Tertiary—about 25 to 55 million years ago—a very important mountain-building event called the Laramide Orogeny took place. This mountain building event was responsible for the Rocky Mountains and affected the area now included in Arches National Park. All of the rocks that were deposited prior to mountain building were folded into anticlines and synclines and were faulted.

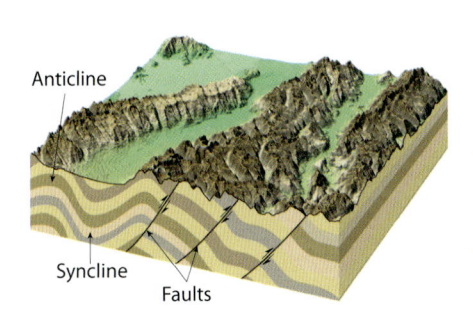

As you travel through the park and the surrounding areas, you will find evidence of the folding and faulting in many of the rocks. The Colorado Plateau began its uplift in the late Tertiary allowing rivers to erode deep canyons. In time, water entering fractures in the rocks filtered down and reached the salt beds and domes. The salt, which is easily dissolved in water, was literally dissolved and eroded underground. The overlying harder sedimentary rocks were suddenly without support from underneath and collapsed.

This is the origin of the long, narrow valley that trends from northwest to southeast—the Salt Valley structure. These structural geologic events eventually contributed to the formation of the nearly 2000 arches seen in the park today.

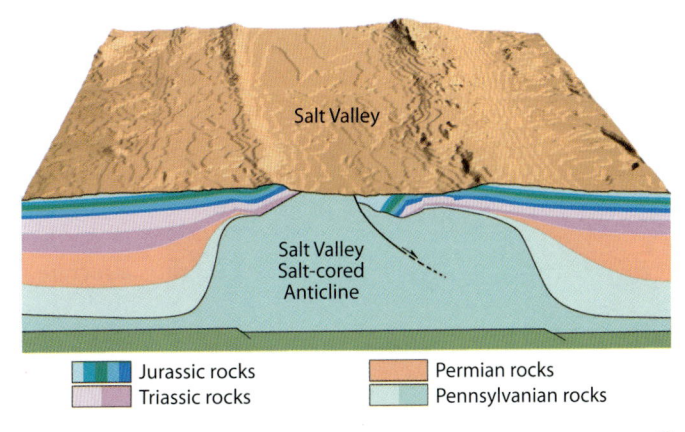

Geologic Map

Maps are invaluable tools not only for geologists, but for anyone who is trying to get from Point A to Point B. Many types of maps exist from simple road maps to topographic maps to the colorful geologic map you see here. Geologic maps show the distribution of rock types at the surface of the Earth as well as the ages of the rocks and the structures found in the rock bodies. Sometimes rocks are covered by recent sediments deposited by rivers, streams, and wind. These loose deposits are usually denoted by yellows and pale orange colors. The most common rocks seen on the geologic map of Arches National Park were deposited during the Mesozoic era and are denoted by the blue colors.

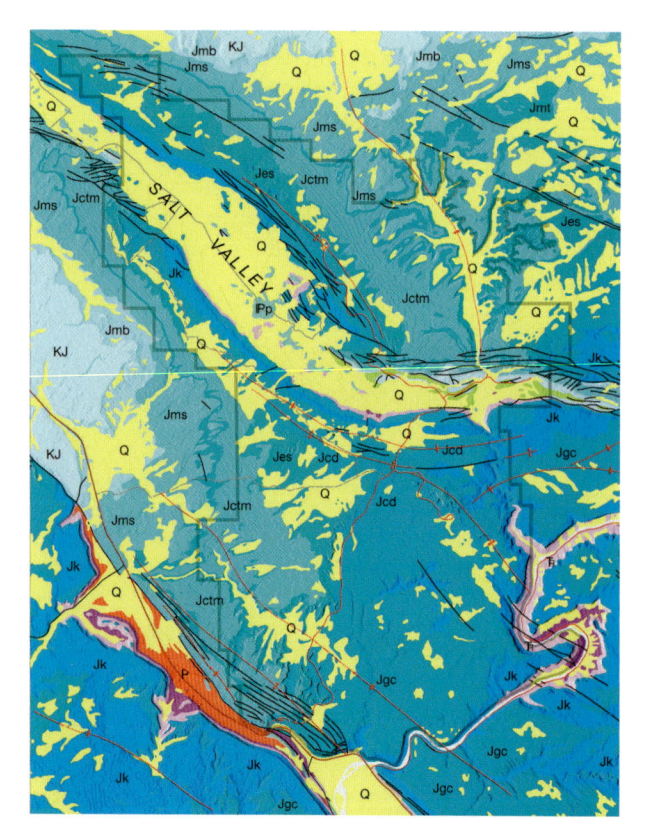

The colored areas on the map also contain abbreviations which help to further identify the formations. For example, the blue area labeled *Jgc* is the Jurassic Glen Canyon Group. The yellow areas labeled *Q* are mostly recent deposits of river-borne sand, silt, and clay.

As geologists walk across the terrain, they locate geologic structures such as folds and faults that occur in the rocks. Several faults and broad, gentle folds span the park. One of the most prominent structures in Arches is Salt Valley, a complex northwest-southeast-trending set of anticlines, synclines, and near-vertical faults that bisect the park. The other prominent structure shown on the map is the Moab Fault system which juxtaposes Mesozoic rocks on the northeast side of the fault against Paleozoic rocks on the southwest side.

Stratigraphic Column

When geologists study the rocks that form the outer crust of the Earth, they divide them into *groups*, *formations*, and *members*. A group is an assemblage of formations that have some geologic characteristics in common; a formation is composed of rocks that are closely related in composition and/or depositional environment. A member is the smallest formal subdivision of a rock unit and shows very distinct characteristics that separate it from other members of its formation.

After a geologist has been in the field for many months (sometimes years), he or she will return to the laboratory, drafting room, and computer and draw a stratigraphic column, a picture of the stack of rocks in the area that was studied. The column is a visual representation of what the geologist saw in the field; special symbols are used to denote the common rock types. Subway tile boxes usually represent limestone, speckled dots represent sandstone, horizontal dashes, shale, and alternating dashes and dots represent siltstone. The irregular edge of the drawing of the rock pile gives a rough idea of how easy it is to weather the rocks. Each rock unit is separated and named (for example, the Paradox Formation or the Entrada Sandstone). These divisions may or may not be subdivided into members depending on how detailed the geologist intends the column to be.

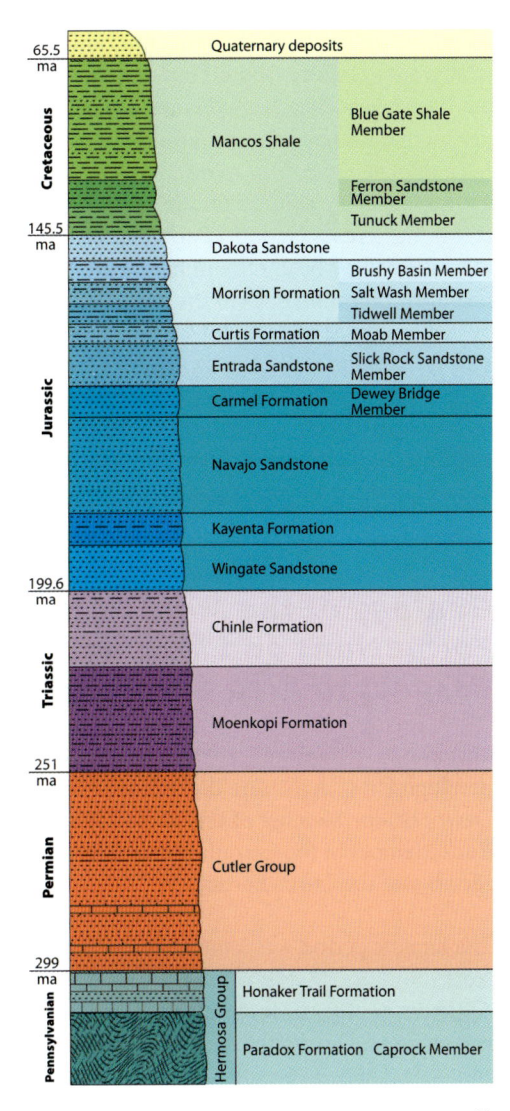

The ages of the rocks are a very important part of the information about the rocks. Each stratigraphic column should not only have the absolute age (that is, how old is the rock in years), but the universally accepted name of the age should be given (for example, Jurassic, a geologic period that spanned about 55 million years).

The stratigraphic column on the previous page shows only the rock formations that are important in the Arches National Park area. Other rocks may be buried deeper under the park and others may have been present before erosion, but the formations shown on this column are the most important ones in the park. Some of the rocks in this stratigraphic column are either buried or eroded and are not visible in the park. The most important formations with respect to arch formation are the Dewey Bridge Member of the Carmel Formation, the Slick Rock Member of the Entrada Sandstone, and the Moab Member of the Curtis Formation. The formations and members on the chart are the currently accepted divisions, but you will find other stratigraphic columns for this area that have different divisions. As geologists continue to study the rocks, ideas of how the formations and members should be catalogued change with time.

Geology of the Most Important Formations and Members with Respect to Arch Formation *Audio track 03*

The four most important rock units that are related to the development of arches are the Jurassic Navajo Sandstone, the Dewey Bridge Member of the Carmel Formation, the Slick Rock Member of the Entrada Sandstone, and the Moab Member of the Curtis Formation.

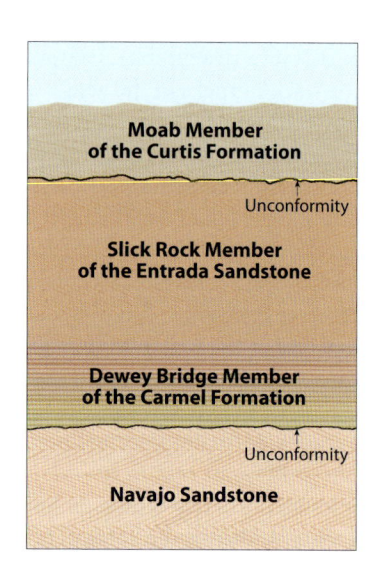

The Jurassic period was a time of emergence over much of North America, that is, the shallow seas that covered the continent at other times in the past were gone and the land was exposed. During the Jurassic, Utah experienced times of complete emergence and times when a small arm of the sea covered part of the area. Along the edges of the sea were tidal flats, beaches, and sand dune fields. Mud, silt, and sand were usually deposited on the tidal flats while dunes and beaches were composed of clean sands.

The Navajo Sandstone was deposited during a dry, emergent time on the North American continent. Vast dune fields, similar to the present Sahara Desert, covered much of the Southwest. As the winds blew, the fine sand collected in large-scale cross-beds in dunes. The sand was eventually buried and cemented with quartz and calcite, then uplifted and exposed at the surface. The ancient dunes are now seen as white to orange cross-bedded sandstones and are called "petrified" dunes, although that adjective is not quite geologically appropriate! Even though the sands were cemented, the sandstones at the surface are friable, that is, they crumble easily in your hand. This is a perfect illustration of the rock cycle where eroded sand is lithified to sandstone which is then, in turn, eroded to sand.

After deposition of the sands that now make up the Navajo Sandstone, a period of nondeposition and erosion took place. This lack of deposition creates an unconformity, a break in the rock record. We know that time is continuous, but we also know that the rock record is not. Sometimes only small fragments of the rock record are missing, but at other times, long intervals pass when no sediment is deposited and much is eroded.

After a period of nondeposition and erosion took place, a span of time that is difficult to determine, the Dewey Bridge Member of the Carmel Formation was deposited. The Dewey Bridge Member is composed of two subunits: a lower yellowish sandstone and an upper reddish-brown silty, shaly sandstone. The thin-bedded, often contorted upper silty, shaly sandstone forms the base on which many of the arches rest. These sands, shales, and silts were deposited near a shallow sea on broad, wet, muddy tidal flats.

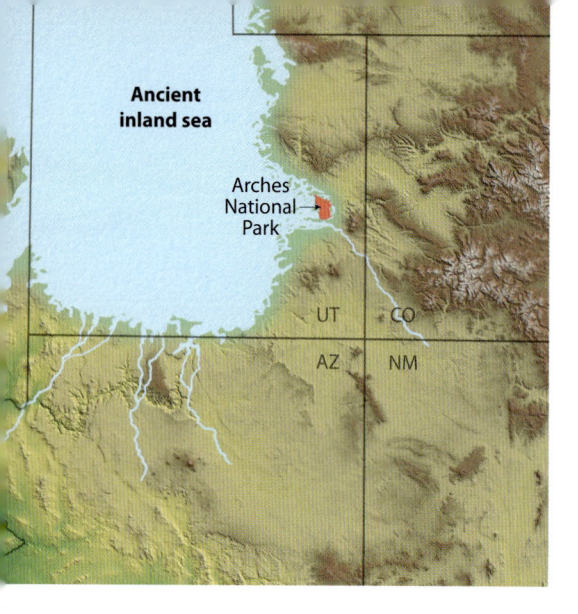

Overlying the Dewey Bridge is the Slick Rock Member of the Entrada Sandstone. The Slick Rock is 200 to 350 feet thick and is composed mostly of a well-consolidated fine-grained sandstone. The sands that make up the formation were deposited in another widespread dune field that may have been close to an inland sea.

The Slick Rock Sandstone is cemented with varying amounts of calcite and iron oxides. Calcite cement is partially responsible for creating the massive, solid rock, but that calcite can also be the downfall of our sandstone. Because rain water is often very slightly acidic, it will slowly dissolve the calcite. When the calcite dissolves, the grains of sand are no longer held together and the sandstone will begin to fall apart. Most of the arches in the park are formed in three subunits of the Slick Rock: the lower unit, the indented middle unit, and the upper unit. The Slick Rock Sandstone varies in color from white to orange-red to brown.

Another unconformity separates the Slick Rock Member from the Moab Member of the Curtis Formation. The Moab sands were probably also deposited by wind. The member forms a weather-resistant sandstone that caps the Entrada. The fine- to medium-grained sandstone is firmly cemented by carbonate cements. In some cases, the Moab forms the upper part of the arch. The unit is variable in color ranging from grey to yellow to orange.

Color of the Rocks *Audio track 04*

The colors of the rocks are one of the most spectacular aspects of the formations in the national parks on the Colorado Plateau, including Arches. Shades of reds, oranges, yellows, and purples change hues as the Sun tracks over the park, sometimes pale in the intense midday Sun, sometimes deep and dark as the Sun sets. These warm colors are the result of very minor amounts of iron compounds entrapped in the rocks between the grains of sand, silt, and clay. If you look at a thin slice of one of these rocks under a microscope, you will see that the grains are usually colorless or white, but each grain is coated with a thin mantle of iron oxide.

Another common coloration seen in the deserts of the Southwest often occurs on the walls of sandstone formations. Dark stains that seem to flow down the sides of the walls are called desert varnish. Desert varnish is formed when wind-blown clay-sized particles stick to wet surfaces on the sandstones. The clays contain manganese oxides that impart black and grey colors and iron oxides that are very dark reds, oranges, and yellows. It is now believed that bacteria play an important part in the retention of the colors by fixing or stabilizing the manganese and iron oxides.

What is an Arch and What is a Natural Bridge? *Audio track 05*

You have probably heard both terms: arch and natural bridge—but did you know that these terms are, geologically speaking, very different? The difference arises from the way in which each is formed. A natural bridge is formed mostly by the action of running water. In other words, streams and rivers are responsible for the erosion of the rock material. A stream may, in fact, still be flowing beneath a natural bridge.

Sipapu Bridge in Natural Bridges National Monument

Photo by Dan Pilkenton

An arch is formed mostly by physical and chemical weathering and mass wasting. Mass wasting is a general term for the downhill movement of rocks, dirt, any Earth material that is in a relatively unstable position. The most important process responsible for mass wasting is gravity. As you explore the arches in the park, you will see how important gravity is in the development of the spectacular features.

Tunnel Arch

Another important part of the definition of an arch is size. Technically speaking, for a structure to be considered an arch, the opening must be complete (that is, you should be able to see through the structure) and the opening must be at least three feet in any direction—top to bottom, side to side, or diagonally.

Most of the arches in the park are formed in the Moab Member of the Curtis Formation, the Slick Rock Member of the Entrada Sandstone, or the Dewey Bridge Member of the Carmel Formation. And of these three, the Slick Rock Member has the most arches. The arches in the Slick Rock Member are positioned along the lower contact with the Dewey Bridge Member, the upper contact with the Moab Member, and in a somewhat weaker zone nearly in the middle of the Slick Rock.

A much less common method of arch formation starts with the formation of a pothole in the sandstone. The exposed surfaces of the sandstone layers are very irregular with small bumps and depressions covering the surface. Water naturally collects in the depressions and in time, may erode the rock to the underlying layer. This is a form of chemical weathering as the cement between the grains is dissolved and the sand grains are washed away.

You may also hear the term "window" as you tour through the arches. A window is essentially still an arch, but some geologists make the distinction that if the opening in the rock is well above ground level, it is a window. If the opening is at or near ground level, the opening is an arch. Unfortunately, when some of the arches and windows were named, these rules were not followed! North and South Windows are actually closer to ground level than Tunnel Arch.

Stages in the Formation of the Arches

As noted in the earlier discussion of the Paradox Basin, the deeply buried salt formations are mobile and can move around in the subsurface. All of this subsurface movement means that the overlying solid rocks will either fold or break. In Arches National Park, salt anticlines and the rocks on top of them are long, narrow upwarps with long axes that trend northwest-southeast. The rocks over the salt anticlines will fold, but because they are more rigid than the salt, they will also break or fracture. These fractures are known as joints. In the sandstones where arches occur, the joints are vertical to sub-vertical and penetrate very deep into the rock formations.

In the following diagrams that illustrate how arches are formed, the 3-D drawings show two formations: the tan-orange, massive top unit represents the well-lithified Slick Rock Member of the Entrada Sandstone. The lower, brown, tan, and white unit that looks like it has a crumbly texture, represents the Dewey Bridge Member of the Carmel Formation. Remember, however, that this arch formation can occur in any hard, fractured rocks that are underlain by softer rocks given the appropriate weathering conditions.

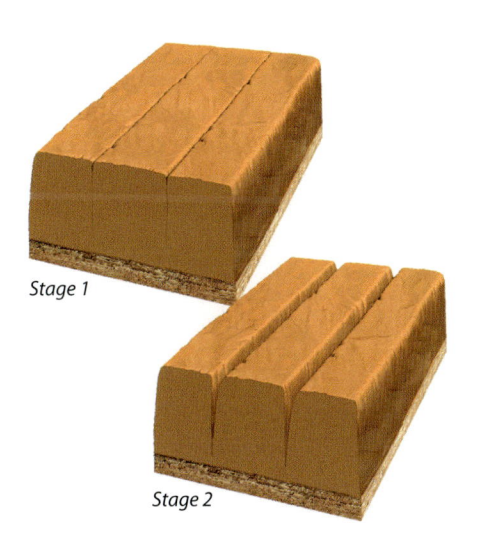

Stage 1

Stage 2

Stage 1. In Stage 1 of arch formation, long, continuous joints form in the hard, brittle sandstones as tectonic processes and salt movement from below lift and fold the rocks. The joints align in one dominant direction (in our case, usually northwest to southeast). The rocks will also have a second direction of fracturing; in our sandstones, that direction is perpendicular, that is 90°, from the first, more prominent set.

Stage 2. The long, thin fractures widen through weathering; the edges at the tops of the joints become more rounded and open as they are exposed to the effects of rain and the freeze-thaw cycle for a longer time. Slowly, the joints reach deeper and deeper into the sandstone.

This fracture in the Slick Rock Member of the Entrada Sandstone illustrates how weathering opens the fractures or joints in brittle rocks. One of the most powerful agents of weathering is water. In this high desert climate, water freezes in the winter and, as we know, water expands when it freezes. This results in freeze-thaw cycles where water enters a small crack, freezes, and expands thereby forcing the rock faces to push away from each other. The water also chemically weathers the rock by dissolving some of the cement, in this case calcite. Now the sharp edges begin to round and the fracture opens even more.

Stage 3. Eventually, weathering and erosion result in a "fin-like" structure as the fractures grow in size both horizontally and vertically. Frost action, exfoliation, and crumbling round off the tops of fins and the joints finally reach to the bottom of the hard sandstone (in this case, the Slick Rock Member) and begin to penetrate the softer underlying shaly sandstone (in this case, the Dewey Bridge Member).

Stage 3

This head-on view of a set of fins shows how the joints in the Slick Rock Sandstone form in parallel rows. Weathering has rounded the tops of the fins and widened the gaps, but the intervening canyons are extremely narrow.

Because the Dewey Bridge Member is less resistant to weathering, the shaly sandstone crumbles and bits and pieces are removed by rainstorms and snowmelt. Notice the fin on the right in the diagram: near the center of the fin, the underlying weathered material begins to move away and curved fractures form in the sandstone.

This cliff face is an excellent illustration of arch formation. A future arch is forming at the contact between the Slick Rock Sandstone and the softer, underlying Dewey Bridge. Exfoliation, or peeling, of the sandstone has resulted in the erosion of thin slabs. As the Dewey Bridge Member is exposed to the effects of rain and wind, the Slick Rock will continue to break away and, eventually, an arch will form.

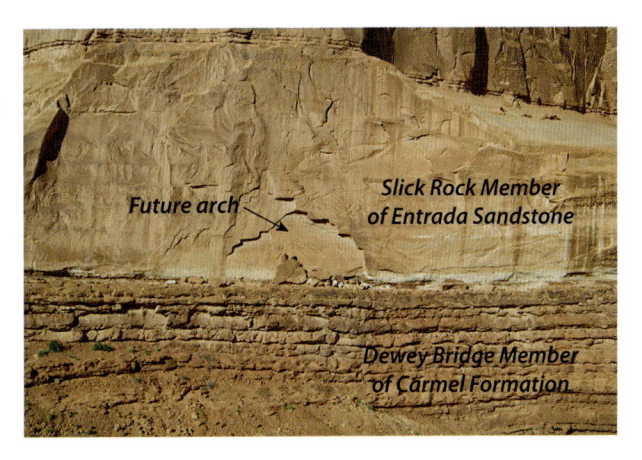

Future arch
Slick Rock Member of Entrada Sandstone
Dewey Bridge Member of Carmel Formation

Stage 4

Stage 5

Stage 4. When the underlying weathered material is removed, a gap or amphitheater begins to form underneath the hard sandstone; now gravity takes over. Vertical fractures that are perpendicular to the original long, narrow fractures become noticeable in the fins as differences in stress fields develop. With nothing to support the overlying strong sandstone in areas of highest stress, arches begin to form. Slabs of sandstone succumb to gravity's pull and begin to fall.

If erosion of the soft bottom layer is slow or does not occur, a fin may simply become narrower and narrower as in the left-most fin or may develop a saddle rather than an arch, as shown in the center fin.

Stage 5. With time and continued weathering, fins progressively narrow, saddles deepen, and true arches form. More and more chunks of sandstone fall into the void and are eventually washed away by flash floods and snowmelt.

Aerial view of fins in Devils Garden

This view of Fin Canyon in Devils Garden illustrates several stages in the formation of arches, the most dramatic of which are the fins. The long, northwest-southeast trending fins in the Slick Rock Member of the Entrada Sandstone form closely-spaced parallel walls of rock. The narrow canyons between the fins are the result of weathering and erosion along joints. Several saddles have formed in the longer walls as erosion sculpts out the sandstone. Notice the many notches in the tops of the fins; these are the second, less prominent set of joints that are perpendicular to the far more conspicuous main joints.

A late stage in the erosion of the fins is illustrated by the outlying pedestals. Perhaps some of these pedestals were connected as arches at some stage in their histories, but now all that remains are groups of pillars, some larger, and some in the final stages of erosion as seen in the bottom left of the air photo.

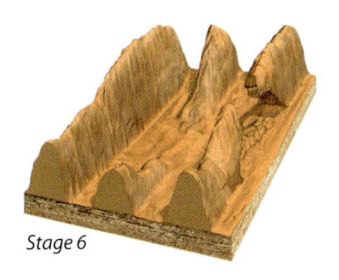

Stage 6

Stage 6. As "old age" approaches for the arch, fractures and gravity conspire to cause the arch's demise. Sandstones may appear to be very strong, but often are very brittle. The weight of the top part of the arch finally exceeds the ability of the legs to uphold the span and the arch collapses. Eventually, unstable fragments at the top will tumble and only the lone abutments will survive—for a time. Spires and spines will endure for many years, but in the end, weathering and erosion will remove all traces of these unique landforms.

Balanced Rock at the entrance to the Windows Section of the park, may also represent the "old age" stage of arch formation. It is possible that at one time Balanced Rock and the adjacent pedestal were connected by an arch. The pile of rocks lying between the pedestals may have connected the two in the past.

Balanced Rock

Wall Arch illustrates that arches do "age" and eventually collapse. After the top span fell in 2008, only the side abutments remain.

Wall Arch

Although the arches are formed by the destruction of material, the arch is one of the most stable structural and architectural forms that a rock formation, or building, can have. All of the rock segments in an arch push in on each other. In geologic terms, all of the stresses are compressional. This means that the rocks in the arch are squeezing against each other and would rather not come apart. Eventually, of course, other processes such as fracturing and gravity will win and the arch will collapse.

All stages of arch formation are present now, were present in the recent past as the Entrada Sandstone was exposed at the surface, and will be present in the future until erosion removes all of the Entrada. Arches may form in other rock formations, but the abundance that we have in Arches National Park in these formations is unique.

TOUR

VISITOR CENTER
Moab Fault
Audio track 06

Arches National Park has some of the most impressive and unique geology in any of the national parks. Although the arches are obviously the geologic structures that everyone comes to see, there are other geologic features throughout the park. One of the first of these structures you will encounter is the Moab Fault.

Very simply, a fault is a fracture surface in the rocks along which the rocks on one side have moved relative to the rocks on the other side. The rocks can either move up and down or side to side or, as is often the case, in a combination of directions. As you read the description of the Moab Fault and compare it to the Park Service interpretive signs, you will notice some differences. Geologists are still studying the exact nature of the Moab Fault, therefore, some differences of opinion will occur. We will try to give you both views on the nature of the fault.

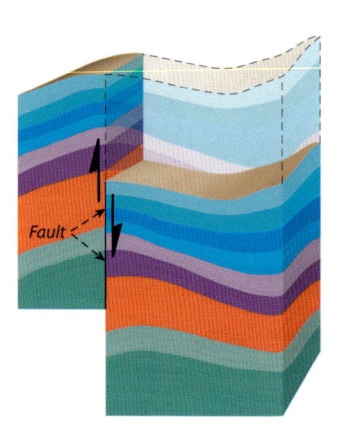

The Moab Fault is considered either a vertical fault as seen on the interpretive sign, or a normal fault. Whether the fault plane is vertical or at an angle, the movement is mostly up and down. The rocks on the southwest side of the fault have moved up relative to the rocks on the northeast side of the fault. In fact, the total displacement is thought to have been 2600 feet. If we put one hand on the rocks on the northeast side of the fault, the rocks on that side were deposited during the Jurassic about 150 million years ago. If we place the other hand directly across on the southwest side of the fault, we are touching rocks that were deposited during the Pennsylvanian, about 299 million years ago.

The complexity of faults can also be problematic. Quite often a fault is actually a zone composed of several faults of various sizes. Such may be the case for the Moab Fault. Instead of one long fault, it may be composed of multiple smaller, interfingering faults.

Regardless of whose interpretation of the Moab Fault is correct, the outcrops along the road are wonderful examples of rock movement and offset of formations. This small fault zone is an excellent illustration of how faulting offsets rock layers. In general, the rock layers to the right have moved down relative to the rock layers to the left. For example, if you could move the block on the right up, you would eventually get Rock Layer 3 on the right to line up with Rock Layer 3 on the left. This is called a normal fault.

Road-cut along U.S. Highway 163/191

But this is not a simple fault. The fault is composed of two primary planes with a thin sliver of rock caught between. In addition, another block on the far left has moved down leaving the center rock unit as the uplifted block with a rather unusual pyramidal shape.

If you continue to look for smaller and smaller features, you will see minor offsets within the larger blocks. These small shifts accommodate all of the minor adjustments as the rocks moved. Notice how some of these small faults fold and pinch the rock, perhaps because the shaly rock is more pliable.

19

COURTHOUSE TOWERS

Queen Nefertiti *Audio track 07*

Along the Park Avenue trail, spires and balanced rocks project skyward from the top of the fins. One of the most notable is the balanced rock that appears to resemble an Egyptian queen. The head of the queen is precariously balanced on a pedestal. Notice that the balanced rock appears to be offset from its base. Geologists suspect that a minor earthquake may have shifted the upper rock just enough to give us the image of Queen Nefertiti, but not quite enough to topple the rock. The next earthquake of any significance in the area may be Queen Nefertiti's downfall!

Park Avenue *Audio track 08*

Park Avenue is a wonderful example of the monumental structures found in the park. The canyon seems to resemble the narrow streets of New York, hence the descriptive name. The narrow fins along the avenue are composed of the sandstones of the Slick Rock Member of the Entrada Sandstone. Spires, notches, and balanced rocks can be found along the tops of the fins.

In this view of one of the walls, a long, laterally continuous, horizontal joint bisects the Slick Rock. A darker orange sandstone overlies a lighter-colored sandstone. This horizontal surface may represent an unconformity, that is, a time period when no deposition took place. The lower sands may have been exposed to weathering and erosion for an unknown length of time before the upper sands were deposited. With time and burial, the sands were lithified, that is, turned into rock, through compaction and cementation. The distinctive lower horizontal fracture probably also represents an unconformity that separates the Slick Rock Member from the underlying Dewey Bridge Member.

Vertical joints seem to be more prominent in the upper, dark orange sandstone. Water seeping into these joints undergoes freeze-thaw cycles accelerating weathering and erosion. The results range from slightly wider, prominent joints, to notches, to isolated spires as more and more material is removed from the wall of rock.

La Sal Mountains Viewpoint Audio track 09

The La Sal Mountains are not located within Arches National Park, but can be seen from many vantage points in the park. The 12,000-foot mountains form a spectacular background, especially when covered with winter snows. At 12,721 feet, Mt. Peale is the highest point in the range.

During the Oligocene and earliest Miocene epochs, approximately 20 to 30 million years ago, igneous activity increased in what is now the eastern Utah area. Hot magma rose from the depths of the Earth and pushed its way into the much older sedimentary rocks that were already in place. These mountains are not volcanoes, but the remnants of the igneous laccolith that intruded into the older rocks. The sedimentary rocks weather more easily, so, through time, the softer sedimentary rocks weathered and eroded leaving the core of the igneous rocks exposed as the La Sal Mountains.

The La Sal Mountains were sculpted by glaciers during the Pleistocene epoch. Although the famous ice sheets of that time did not cover Utah, episodes of colder climate caused the expansion of alpine glaciers in many parts of North America, including the Utah area. The La Sal Mountains may have suffered at least nine glacial advances. Erosion caused by alpine glaciation results in rugged topographic features that give us the spectacular scenery of many of the mountain chains in the western United States.

Courthouse Towers Viewpoint *Audio track 10*

The Courthouse Towers provide a more blocky, monolithic expression of fin development. Visible in this view are the same features seen at Park Avenue. Horizontal joints formed along probable unconformities separate the Dewey Bridge from the Slick Rock and divide the Slick Rock into two subunits. The darker upper unit probably contains slightly more iron giving it the rich orange coloration. The lower Slick Rock subunit has less iron and hence, the color is less pronounced.

Another feature very common in the deserts of the Southwest is seen on the walls of the monoliths. Dark stains that seem to flow down the sides of the walls are called desert varnish. Desert varnish is formed when wind-blown clays stick to wet surfaces on the sandstones. The clays contain manganese oxides and iron oxides that form blacks, greys, and dark reds and yellows. Bacteria fix the manganese and iron oxides resulting in the dark stains.

Three Gossips
Audio track 11

The Three Gossips are a group of spires with balanced rocks on top of each. The balanced rocks have been sculpted by the freeze-thaw action of water and by wind and gravity to resemble the heads of three people passing time talking with each other. The darker upper Slick Rock Member composes the bodies and heads of the Gossips as they stand on the lower, lighter-colored Slick Rock Member.

Sheep Rock *Audio track 12*

Sheep Rock, standing alone at some distance from the wall to its left, looks like a lonely sheep who has lost his flock. But what if the sheep had not always been separated from his flock? The second image of sheep rock provides us with a hypothetical reconstruction of what this formation may have looked like in the past. The sheep would have formed the abutment of one smaller arch with the top span coming from his head. This smaller arch would have rested on the pedestal that is still seen today between the sheep and the wall. A second larger arch would have connected the pedestal to the wall on the left. If we could step back one stage farther in time, the entire wall would have been intact forming a fin! This is all hypothetical, but the pieces of rock that are still in place suggest that the hypothesis is reasonable.

Petrified Dunes

THE GREAT WALL
Petrified Dunes Viewpoint Audio track 13

The Navajo Sandstone which underlies the Dewey Bridge Member of the Carmel Formation is one of the most widespread formations in the Southwest. The Navajo Sandstone is also Jurassic in age, approximately 170 million years and older. Wind-blown sand dunes covered much of the Southwest during this time in a vast area similar to our present-day Sahara Desert. These sand dunes were eventually buried and cemented with calcite and quartz turning the dunes into sandstone. As the Colorado Plateau uplifted and overlying rocks eroded, the Navajo Sandstone was exposed. Now these large-scale cross-beds are seen in many places in the Southwest. Here, the lithified dunes are light in color, in other words, the rocks contain very little iron oxide.

The petrified dunes are starting to erode; the wind-blown sand from the sandstones is once again accumulating in dunes. This is an excellent illustration of the rock cycle as sand is reworked over and over again lithifying into sandstone and sandstone weathering into sand.

As you continue driving north on the main road, the Great Wall is on your left. This wall of rock is composed of the Entrada Sandstone and rises to almost 300 feet in places.

The Great Wall   Photo by Dan Pilkenton

THE WINDOWS SECTION
Balanced Rock Audio track 14

Balanced Rock is one of the most photographed structures in the park. Located near the entrance to the Windows Section of the park, the balanced rock on its pedestal stands 128 feet tall. The balanced chunk of Slick Rock Sandstone on the top is 55 feet tall and estimated to weigh more than 3500 tons. By definition, a balanced rock is precariously posed on a pedestal base, ultimately awaiting the effects of gravity to bring it tumbling down.

Note the soft sediment deformation of the Dewey Bridge Member (the lower, dark red formation) in the pedestal to the right of Balanced Rock. Over 150 million years ago as the Slick Rock sands were deposited, the weight of the sands pressed into the soft, muddy sediment of the Dewey Bridge resulting in the contorted bedding that later lithified into the rock formation that we see today.

Balanced Rock may represent the "old age" stage of arch formation. It is possible that at one time Balanced Rock and the adjacent pedestal were connected by an arch. The pile of rocks lying between the pedestals may have connected the two in the past.

Balanced rocks, are by definition, ephemeral features. A third pinnacle located near the Balanced Rock and known as Chip off the Old Block fell during the winter of 1975/1976.

28

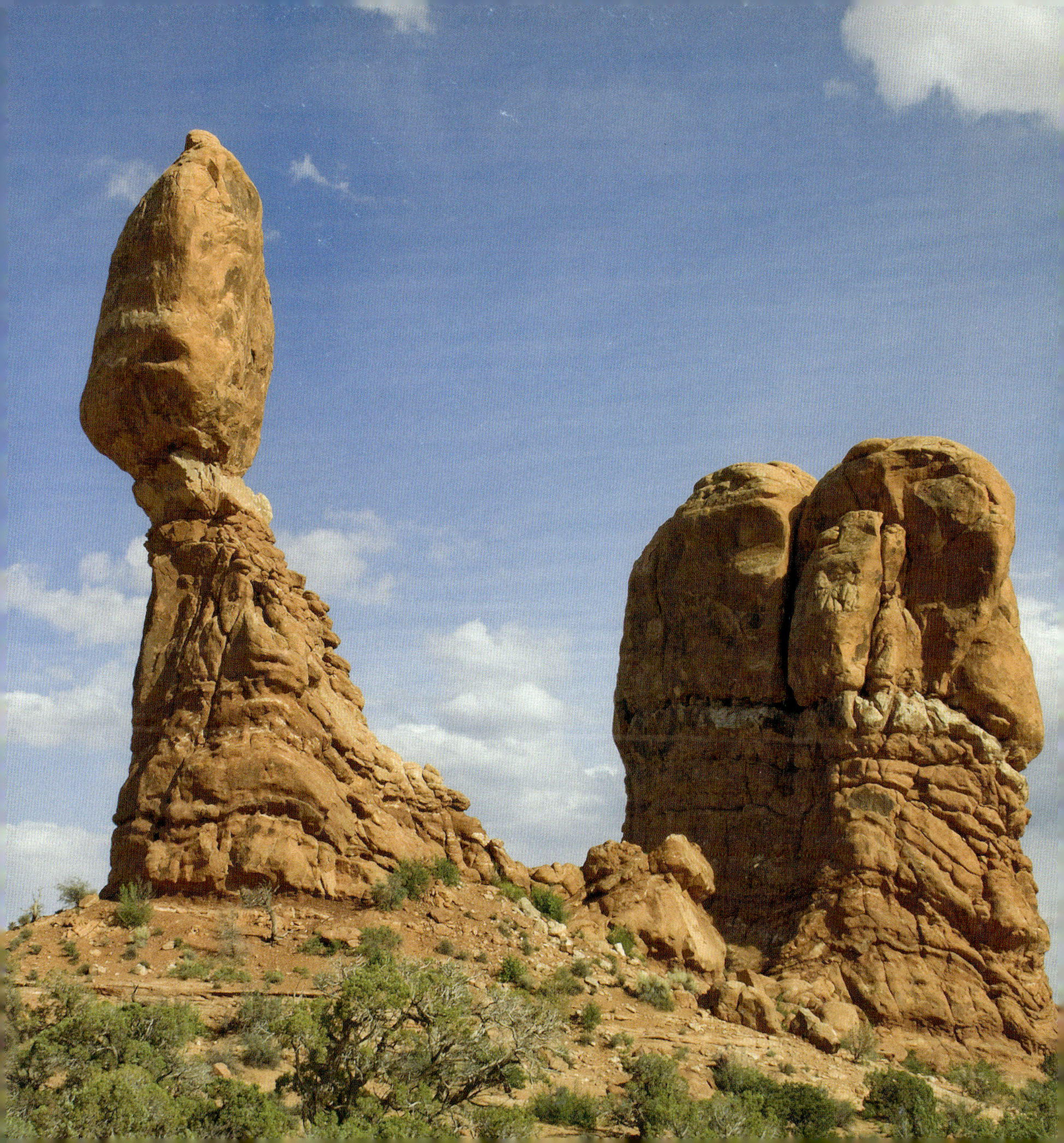

Garden of Eden *Audio track 15*

The Garden of Eden in the Windows Section is composed of numerous pinnacles with a few small windows and arches. Joints separate the rocks into closely-spaced towers. Weathering has contributed to the rounded appearance of the sandstones and siltstones. In the image on the facing page, a rock climber descends the highly contorted Dewey Bridge Member; his climbing partner waits for him near the Dewey Bridge–Navajo Sandstone contact at the base of the tower. Some of these towers have thin caps of Slick Rock Sandstone.

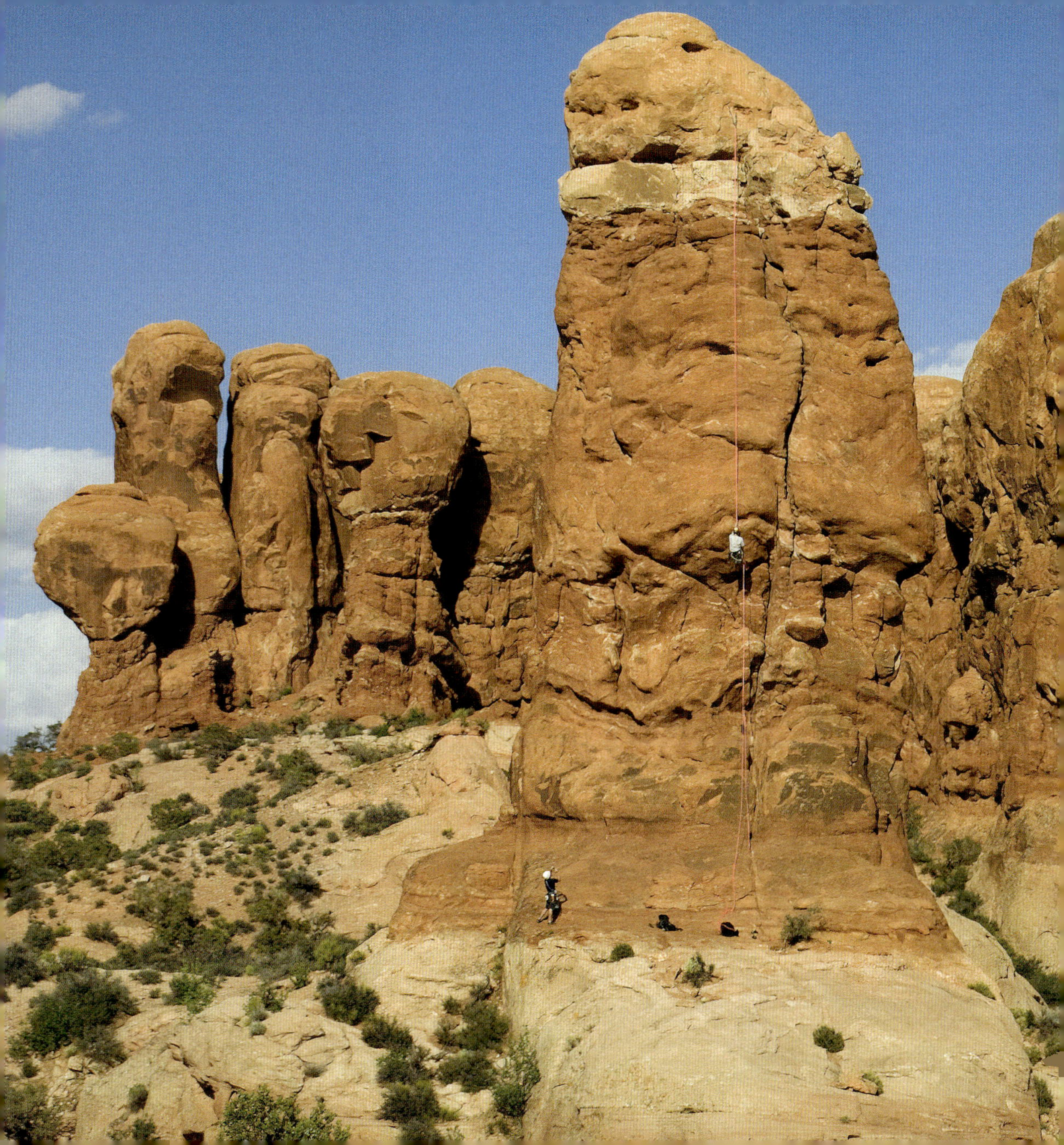

Parade of Elephants *Audio track 16*

This fanciful stone structure is called Parade of Elephants for good reason. Mother Nature has used the power of weathering to sculpt this massive section of the Slick Rock Sandstone into fanciful elephantine forms. Horizontal fractures in the bulges along the sides resemble trunks and the broad domes at the tops, the heads of the elephants. How many elephants do you see parading across the landscape?

The underlying Dewey Bridge Member is visibly contorted here as though the weight of the elephants has squeezed footprints into the shaly, silty sandstone. Geologically speaking, of course, this is not quite correct! It is readily apparent that the Slick Rock Sandstone is very massive. As the Slick Rock Sandstone was deposited on top of the muddy, silty, sandstone of the Dewey Bridge, the lower unit was contorted. This happened very soon after deposition when the Dewey Bridge sediments were still soft and full of water. The weight of the sand deposited on the top squeezed the softer sediment below resulting in the wavy contortions in the Dewey Bridge found in several of the arches in the park.

The development of arches and alcoves in the mid-sections of the elephants does not detract from the illusion. An opening that diminishes in size cross-cuts the entire wall. The larger section of the opening began as an alcove, that is, the sandstone shed rock material creating an arch-like form, but the gap did not open through the entire arch. With continued erosion, alcoves may eventually form arches.

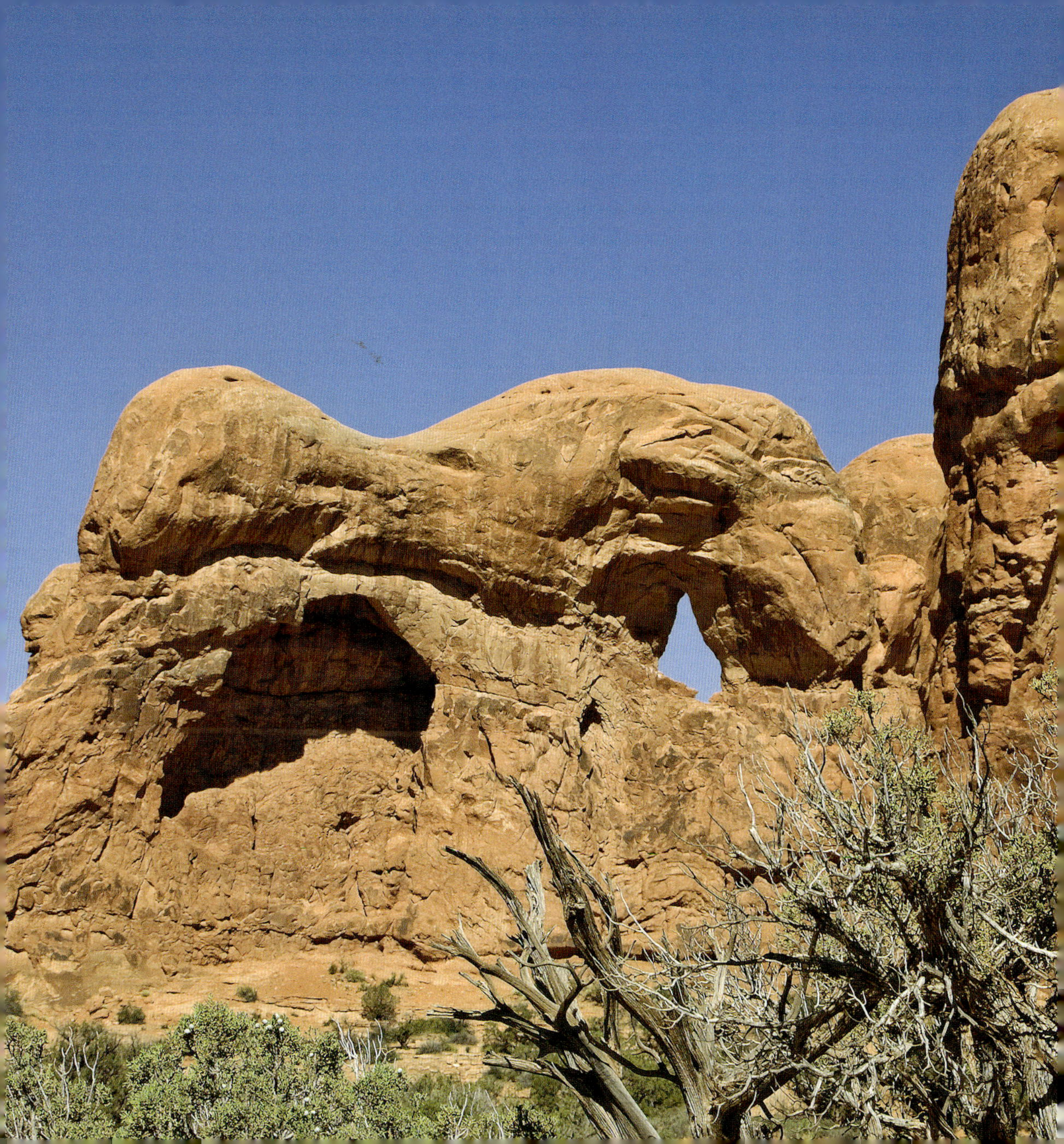

Double Arch *Audio track 17*

Double Arch is an unusual pair of arches that are quite close together. The southeast arch is the second largest arch in the park; it is 105 feet high and spans 160 feet at its widest (notice the people in the photo for scale). The west arch is smaller with a height of 61 feet and a width of 60 feet. The arches appear to be nearly at right angles suggesting that one formed in the primary joint direction and the other formed in the secondary joint direction. This pair is typical of the arches in the Windows Section as they are formed in the lower Slick Rock Member of the Entrada Sandstone and rest on the softer, more easily erodable Dewey Bridge Member of the Carmel Formation.

The origin of Double Arch is thought to be somewhat different than most in the park. Rather than erosion of the Dewey Bridge Member and collapse of the Slick Rock Sandstone by gravity, this pair of arches may have formed by water erosion from above. Water collecting on the upper surface of the exposed rock in a pothole would have eventually chemically removed the calcite bit by bit. Holes would erode through the sandstone eventually forming the Double Arch, aptly known as a pothole arch.

Gravity does play a role in the continuing development of the arches. The top of the larger arch shows signs of thinning; horizontal fractures are becoming more pronounced at the apex of the arch. The smaller west arch has a very thick crown, but notice the rectangular gap on the top span; a relatively large flat block is missing. The very angular edges and fresh surfaces suggest that the block fell in the not too distant past.

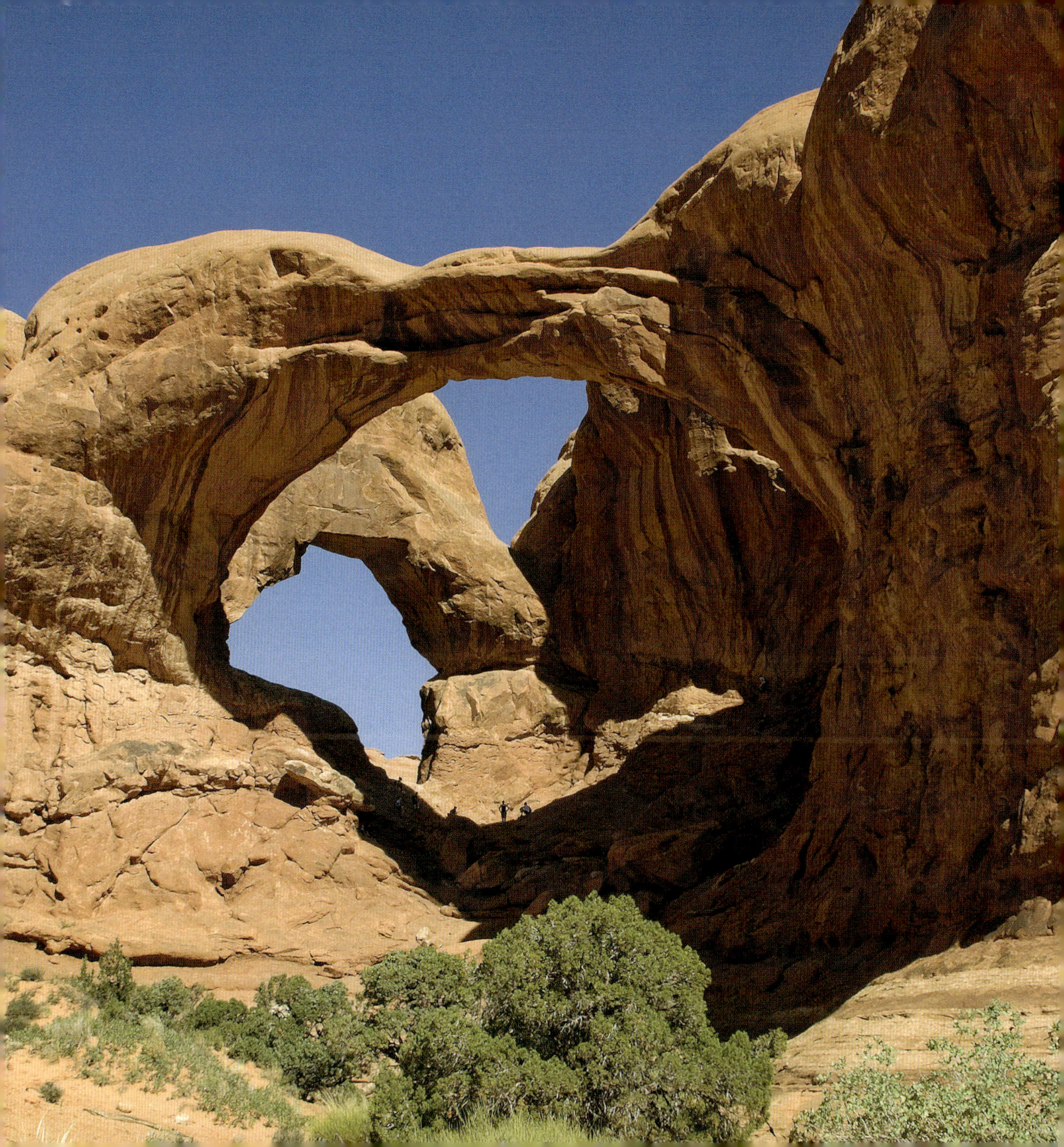

North Window

North and South Windows *Audio track 18*

North Window and South Window are a pair of well-formed arches standing side by side. These are cliff-wall arches as they are embedded in a fin rather than free-standing. If we were to classify them in our Stages of arch formation, they would be classified as mature, or Stage 5 (page 13). The wavy-bedded Dewey Bridge muddy siltstone forms the base of the arches; each arch is within the Slick Rock Member of the Entrada Sandstone. If you stand back from the twin arches, some people think that the protruding rock seen between the arches looks like a nose and the arches a pair of eyeglasses!

South Window — Photo by Dan Pilkenton

Turret Arch Audio track 19

Turret Arch is appropriately named for its adjacent tower. A turret is a small, tower-shaped projection on a building. The Slick Rock Sandstone tower projects not from a building, but from the side of the arch structure. In time, the turret may weather to form a balanced rock with the arch at its side.

The irregular arch has an opening that is almost twice as high as it is wide. The arch is 65 feet high and only 35 feet wide. A second arch, or small window is forming in the wider arm of the arch.

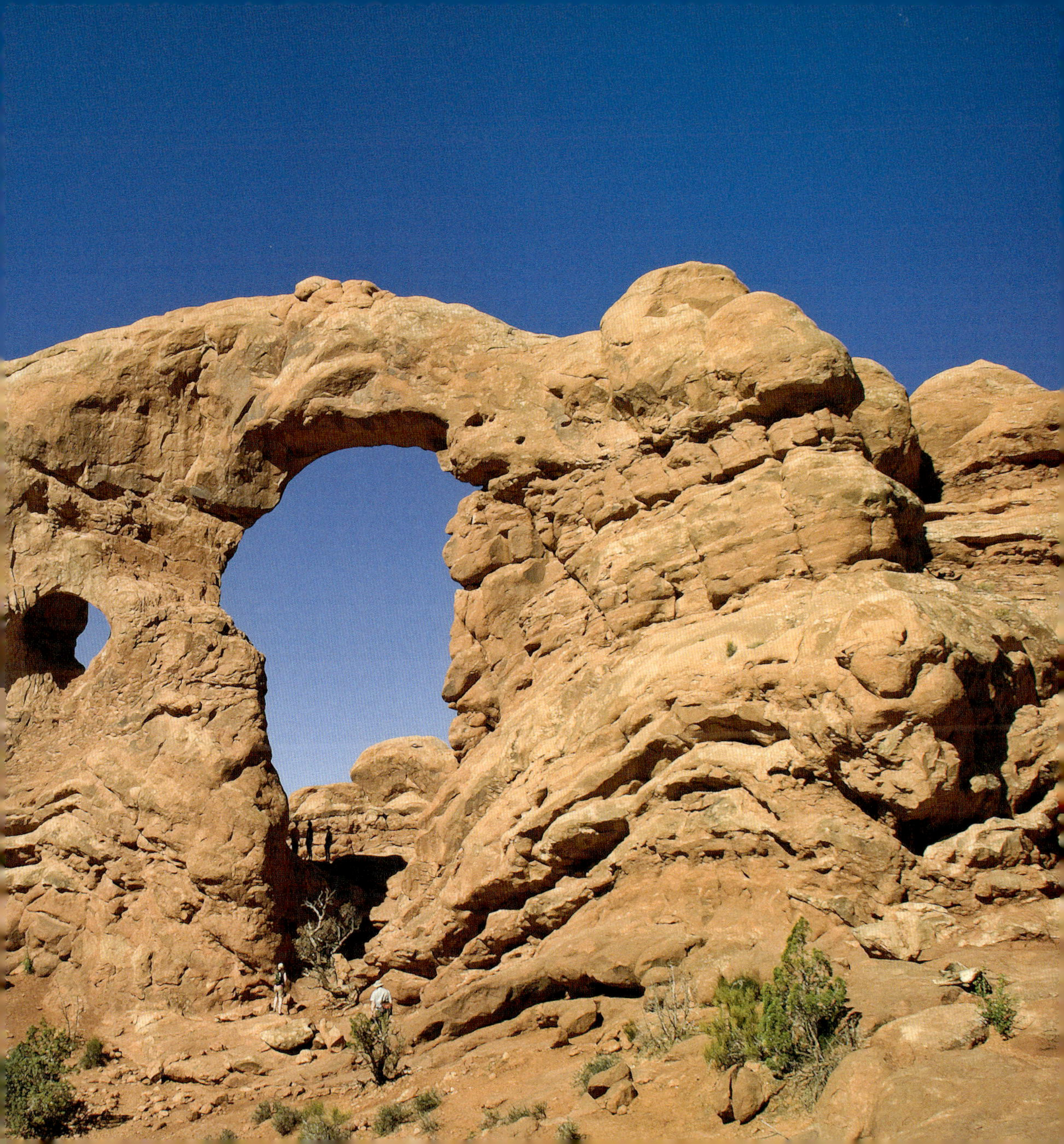

WOLFE RANCH
Delicate Arch Audio track 20

Delicate Arch is, perhaps, the most famous arch in the park. You will see drawings, paintings, and photographs of the arch everywhere from television ads for the state of Utah to Utah license plates!

The one and a half-mile walk to Delicate Arch is rather strenuous, but the world-famous view is worth the effort. (Be sure to carry adequate water, especially in the extremely hot summer months!) After climbing over rolling hills covered with vegetation, you will reach the slickrock exposure of the Entrada Sandstone. The remaining hike is a 500-foot climb on the sandstone. As you reach the top, you will see one of the most beautiful views in the park—Delicate Arch. The arch, with the La Sal Mountains in the background is a favorite photo opportunity for tourists, journalists, and illustrators.

The arch is 45 feet high and the opening is 33 feet at its widest; the uppermost part of the arch is 19 feet thick. Delicate Arch is a free-standing arch, that is, its side abutments are no longer attached to fins. The arch is formed in the Slick Rock Sandstone Member of the Entrada Formation with the Moab Member of the Curtis Formation forming a thin (only a few feet) cap on top of the arch. Notice how the differential erosion of the sandstones comprising the arch delineates some of the bedding planes within each formation. Slightly softer laminations are eroded just a little more than slightly harder laminations leaving thin indentations along bedding planes. A prominent erosional indentation is apparent on the narrower leg of the arch about one-third of the way up from the bottom. This is the narrowest part of the arch at only five feet.

Vertical joints are also common in the arches. Two are seen in the lower part of the thinner leg and at the base of the thicker leg. These vertical cracks lend some instability to the arch and, in the end, may result in the downfall of Delicate Arch.

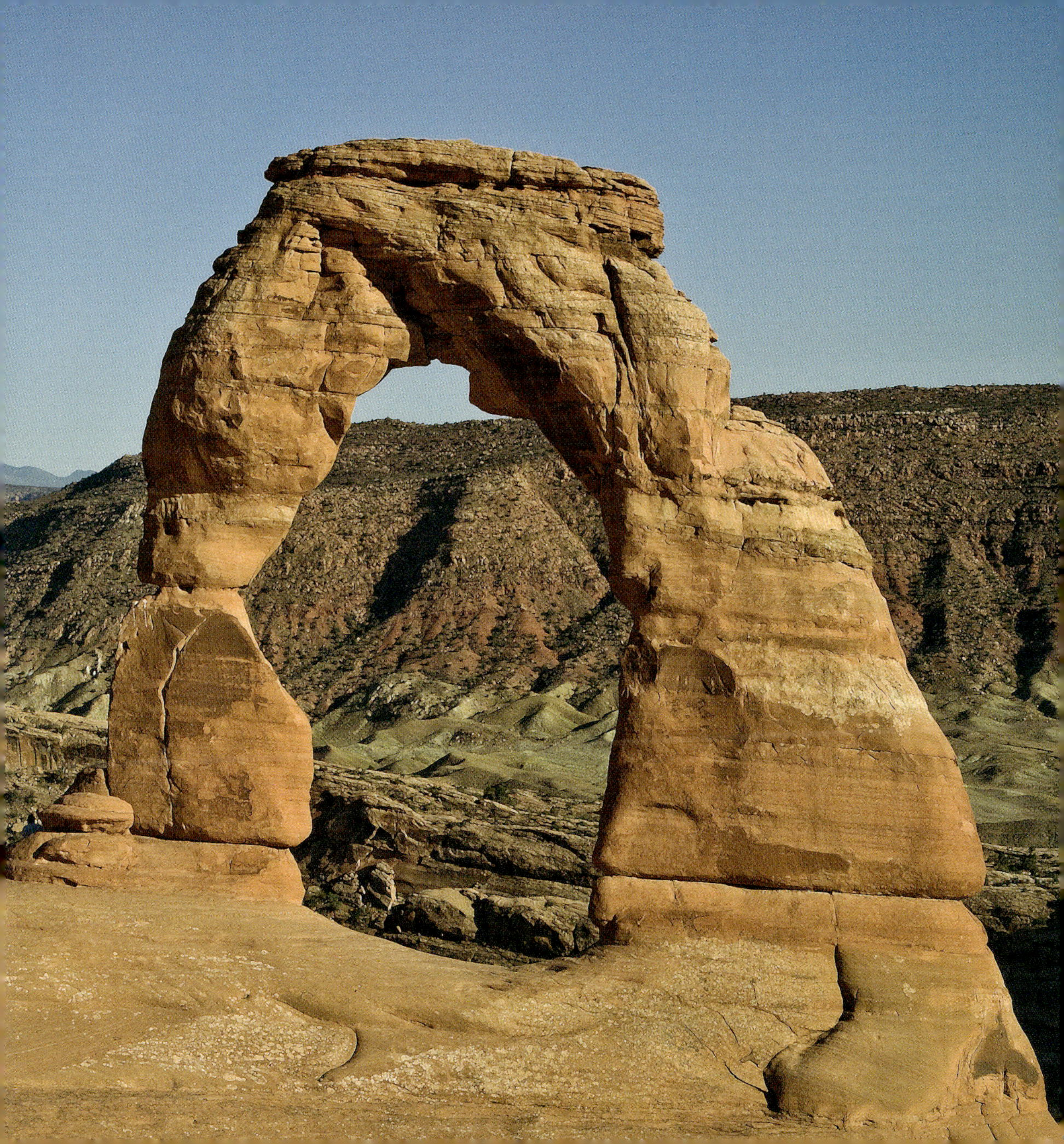

Delicate Arch Viewpoint *Audio track 21*

Notice as you walk around Delicate Arch your view changes from the longer distances of the La Sal Mountains to closer pinnacles and spires near the arch. The constant near-horizontal nature of the Entrada Sandstone layers is apparent. The "mounds" of rock in the background are remnants of fins and possibly, remnants of arches that have fallen. The distinctive indentation near the base of the arch legs is easily traceable from Delicate Arch to the rock forms in the background suggesting an aerially extensive thin rock layer that was softer or not as well cemented as most of the other rock.

As your gaze moves down, two other geologic features become apparent: small holes are beginning to form in the steep cliffs of the sandstone. These holes are called tafoni, or informally, stonepecker holes in joking reference to a hypothetical bird that would be crazy enough to try to peck holes in sandstone! But that is not how these holes are formed. Small pieces of sandstone chip out of the rock face through freeze-thaw action and gravity. In many cases, these tiny caves are aligned along certain beds suggesting softer and less well-cemented rock. Through time and the erosive power of water and freeze-thaw action, the tafoni enlarge.

The second feature visible in the massive sandstone units is the presence of cross-bedding. These laminations within beds tell us something about the original depositional environments of the sands that eventually became sandstones. The sands that comprise the Entrada Formation were deposited during a very dry period of the Jurassic period. Large sand dunes covered much of the Utah area during this time. As the sand was blown across the arid landscape, dunes would form just as they do in modern deserts. These cross-beds are the preserved windward and leeward internal laminations of dunes as the sand was blown across the vast desert.

FIERY FURNACE
Fiery Furnace Viewpoint *Audio track 22*

Fiery Furnace is an outstanding illustration of joint and fin development in the Slick Rock Member. As the salt dome or salt wall beneath what is now the Salt Valley arched upward forming the Salt Valley Anticline, stresses on the sandstone caused incipient joint formation. As the salt dissolved and the anticline collapsed, the joints were accentuated.

Fiery Furnace aerial view

In the white sandstone on the left of the aerial photo, most joints penetrate only a short distance downward. Because the joints are now long, narrow depressions, water collects in them providing just enough moisture for vegetation to take root in the weathered sand. Notice that in the foreground, the long joints are parallel; in the background, some joints bifurcate and intersect. It is also evident that numerous secondary joints crosscut each rounded future fin at right angles.

As you look to the right of this photo, the joints in the red sandstone have been widened by chemical and physical weathering and the detritus removed by erosion. The slot canyons developed between the fins in the Fiery Furnace are so narrow that it can be difficult to squeeze through the openings.

As rain and snowmelt run down the long joints, the water will collect at low points. Two intermittent streams have eroded washes perpendicular to the fins. The mouths of the washes are V-shaped and widen to the right. As the channeled water exits the fins, the flow rate decreases and sediment is dropped in a fan-like shape.

Fiery Furnace

This view of the Fiery Furnace shows the jumble of fins and pedestals at the edge. The Furnace was so-named because of the spectacular orange-red glows in the slot canyons as the Sun moves across the rocks. We can see in this view the slight tilt of the sandstone beds associated with a rollover anticline associated with the Salt Valley Anticline. The pedestals and fins form such a labyrinth that very experienced rangers must guide you through the maze.

Sand Dune Arch *Audio track 23*

As you stand in front of Sand Dune Arch you can almost imagine that you are standing on Luke Skywalker's home world of Tatooine. In this view, we see only rock and sand. The small arch, just 12 feet high, has a joint bisecting the top of the arch. Although this will eventually weather and open, the abutments are so thick and stocky that the roof may remain attached even if it separates into two unequal parts. Blocks of rock from the arch litter the ground underneath. Evidence of exfoliation is present on all of the rock faces. The massive rock that forms the left abutment has a line of tafoni in the second indent from the top.

Look closely at the rocks that form the arch and walls. You can see the original cross-bedding of the Slick Rock Member of the Entrada Sandstone. The sandstones are now weathering to produce the sand that covers the trail.

Fins near Sand Dune Arch Audio track 24

From the ground, the sandstone fins are imposing structures. The intense red-orange color of the Slick Rock Sandstone stands out sharply against the blue of the sky and the green of the vegetation. As discussed in the Introduction, the oranges, reds, and yellows result from minor amounts of iron contained in the sandstone. The lighter shades usually mean less iron; as the rock beds approach white, iron is lacking entirely. As you face the edges of the fins, it is easy to see why such descriptive names as "Wall Street" and "Park Avenue" have been given to the fins and pedestals in the park.

In this view, you are looking directly into the primary joints which separate each fin. In some cases, the joints have weathered to the ground surface forming steep-walled, narrow slot canyons. Other joints extend to the ground, but weathering has not yet opened the space enough to allow entrance by anyone except small desert creatures and plants. Notice how the edges of the fins have been rounded by weathering as rain, the freeze-thaw cycle, and wind have plucked the grains of sand from the stone.

DEVILS GARDEN
Skyline Arch *Audio track 25*

Skyline Arch is one of the arches that may be better described as a "window." The opening is relatively high in the cliff face rather than closer to the ground. The low angle dip of the sandstone is apparent here as are the joints perpendicular to the bedding. As weathering continues, these joints will widen and become towers adjacent to the arch. That is, of course, if the arch is still standing! Notice the sharp-angled section of the arch on the left (northwest). A large block of rock fell from the top of the arch in 1940, almost doubling the size of the opening. The rubble from the fall is still lying at the base of the arch.

Aerial Fins in Devils Garden

In addition to the rugged beauty of eastern Utah, this view of Fin Canyon in Devils Garden illustrates several stages in the formation of arches, the most dramatic of which are the fins. The long, northwest-southeast trending fins in the Slick Rock Member of the Entrada Sandstone form closely-spaced parallel walls of rock. The narrow canyons between the fins are the result of weathering and erosion along joints. Several saddles have formed in the longer walls as erosion sculpts out the sandstone. Notice the many notches in the tops of the fins; these are formed in the second, less prominent set of joints that are perpendicular to the far more conspicuous main joints.

A late stage in the erosion of the fins is illustrated by the outlying pedestals. Perhaps some of these pedestals were connected as arches in the past, but now all that remains are groups of pillars, some larger, and some in the final stages of erosion as seen in the bottom left of the air photo.

Tunnel Arch *Audio track 26*

Following the nomenclature of window and arch, the Tunnel Arch may be considered a window as it is above ground level. Whichever designation we choose, the name "tunnel" is appropriate. The opening is nearly circular with a height of 26 feet and a width of 24 feet; the thickness of the walls of each abutment may be greater than 10 feet.

The arch has been formed completely within the massive sandstone of the Slick Rock Member. The darker orange-red sandstone forms the thick cap of the arch while the base is resting on a white, massive sandstone. Desert varnish stains the upper part of the arch. Higher in the orange-red sandstone another smaller circular arch, or window, has formed.

Pine Tree Arch *Audio track 27*

Pine Tree Arch is named for the pinon pine tree that is in the foreground of the photo. The tree is perfectly framed by the arch from the other side. The large angular gap at the top of the opening suggests a fairly recent rock fall. The edges of the gap have not yet been rounded by weathering. Vertical joints on the abutments show considerable rounding resulting from freeze-thaw action, water flowing over the surface, and wind abrasion.

Landscape Arch Audio track 28

Landscape Arch is not only the longest span arch in Arches Park, but one of the longest arches in the world. Using laser technology, Jay Wilbur of the Natural Arch and Bridge Society, accurately measured the interior span of Landscape Arch to determine if this arch or Kolob Arch in Zion National Park was larger. After painstaking measurements, Wilbur determined that Landscape Arch is 291 feet across the interior opening making it about three feet larger than the Zion arch. This not only makes Landscape Arch the largest in the United States, but possibly the world.

It is apparent that Landscape Arch is fragile; the upper part of the span is extremely thin. At most, the arch is only 6 feet thick. Thin slabs are peeling, or exfoliating, from what remains of the arch. Three large slabs from 30 feet to 70 feet have fallen since 1991. The angular, relatively unweathered appearance of the abutments suggest recent rock falls. Most of the rubble under the arch is from the multiple episodes of collapse. Unfortunately, this delicate arch may not survive too many more years. The arch is now considered so dangerous that the trail beneath it has been closed.

The aerial view of the arch shows an undulating surface attributed to slope wash as debris from the uphill area becomes saturated with rain and snowmelt, then slumps downhill under the arch. Landscape Arch differs from many of the other arches in the park as it is not as orange-red indicating that the sandstone does not contain as much iron oxide in this area as other areas in the park.

Landscape Arch aerial view

Wall Arch
Audio track 29

Wall Arch is a dramatic illustration of Earth processes in action. This 12th largest arch in the park was one of the most photographed on Devils Garden trail. Sometime during the night of August 4th, 2008, gravity won the final battle. The apex of the arch had thinned. The entire arch structure had many joints. Geologists think that a very minor earthquake may have been the last straw for Wall Arch. A minor shake overcame the strength of the rocks to hold the arch form and the free-standing arch collapsed. The trail under the arch has been diverted as all of the rubble fell on the trail. Large chunks of rock on the remaining arms are still unstable and will continue to fall. How fortunate that no one was standing underneath at the time of the fall!

*Wall Arch **before** August 4th, 2008*   *National Park Service photo*

*Wall Arch **after** August 4th, 2008*   *National Park Service photo*

Wall Arch viewed from today's trail

Partition Arch
Audio track 30

Partition Arch has a rather unusual feature. The arch appears to be a single opening, but just off-center is a pedestal. The pedestal almost appears molded as though it is holding up the arch. Geologically, the pedestal may help stabilize the arch giving the arch more strength to hold its massive roof.

Partition Arch is also a fine example of desert varnish. The black and brown stains are prominent on the upper and lower ledges. A bedding plane in the sandstone probably provided a horizontal parting for water to enter and erode the notch.

Partition Arch

Double O Arch *Audio track 31*

Double O Arch is a very unusual arch formation. Although pairs of arches are commonly found side by side, it is rare to find one arch on top of another, especially so close together. Both arches are in the Slick Rock Sandstone. The larger top arch is approximately 35 feet high and 67 feet wide. The relatively flat top on the upper arch shows few joints, but seems rather tenuously attached to the main sections of the fins.

Double O Arch

The lower arch is much smaller at 9 feet high and 21 feet wide. Notice the angled slab in the ceiling of the lower arch. With continued freeze-thaw, this slab will probably soon be down.

A small tafoni is forming to the right of the upper arch. In time, this small hole may become an arch. Wouldn't it be extraordinary if three arches formed in a triangular pattern? Unfortunately, the lifetimes of the two existing arches will probably be long over before that small notch becomes an arch. In fact, as the tafoni increases in size, it will probably intersect the upper arch and possibly merge with it.

The fin abutment on the left shows the typical signs of arch formation as slabs of rock have fallen. Unfortunately, continued arch development in this area would mean the collapse of the existing upper arch.

Glossary

Aeolian processes: erosion, transport, and deposition of material by the wind.
Alcove: an incipient cave in a cliff face. With time, weathering, and erosion, an alcove may become a cave, and eventually, an arch.
Alpine glacier (mountain glacier, valley glacier): long, ribbon-like glacier confined to mountain valleys.
Anticline: a type of fold that is convex upward.
Arch: in nature, an opening in a cliff formed mostly by physical and chemical weathering and mass wasting (gravity). Technically, a natural arch must have an opening that is complete (that is, you should be able to see through the structure) and the opening must be at least three feet in any direction. See free-standing arch, cliff-wall arch.
Bedding plane: a planar surface that separates strata.
Carbonate: a group of minerals uniting a variety of cations with the CO_3^{-2} anion. Members of this group of minerals include calcite, dolomite, aragonite, siderite, and others.
Cement: minerals that fill or line pores in rocks.
Cliff-wall arch: an arch that is located in a cliff or fin whose abutments are not discrete (for example, North and South Windows).
Cross-bed: sedimentary structure in which sets of layers of sediment are inclined with respect to each other. These can be formed in water or by deposition of sand and silt in aeolian environments.
Erosion: movement of soil and rock material by water, wind, and gravity.
Exfoliation: separation of the surface layers of a rock that resembles the layers of an onion that are peeled. The separation may start as pressure release when a rock is brought to the surface, then may be enhanced by chemical and physical weathering.
Fault: a planar surface defined by a break in rock formations that shows movement of the rock on one side relative to the rocks on the other side. Faults are classified as normal (extensional stress fields), reverse (compressional stress fields), and strike-slip (shear stress fields).
Fold: bend in the rock strata; see anticline, syncline.
Fracture: a break in a rock.
Free-standing arch: an arch whose abutments stand alone resulting in an arch that appears isolated from surrounding cliffs (for example, Delicate Arch).
Friable: a rock or soil that crumbles relatively easily with hand-applied pressure.
Igneous rocks (intrusive and extrusive): one of the three main rock types, igneous rocks are crystallized from molten magma. In the simplest definition, the magma may cool deep within the Earth resulting in igneous rocks like granite with larger crystals (intrusive igneous rocks) or the magma may flow out on the surface of the Earth forming lava flows that have microscopic crystals (for example, basalt) or explode violently, such as Mt. St. Helens (extrusive igneous rocks).
Indurate: to harden; geologically, the process of turning sediment into rock by compaction and cementation (sedimentary rocks) or cooling of a magma body (igneous rocks) or stabilizing metamorphic rocks.
Joint: a fracture in a rock that shows no appreciable movement. That is, the rocks have not moved up or down or side to side with respect to each other.

Laccolith: a type of igneous pluton that appears lenticular or circular when seen from above, but in cross-section has a flat floor and domed roof. The term laccolith is an older term that has fallen into disuse in favor of other descriptive terminology.
Lamination: very thin stratification, usually less than 10 millimeters in thickness.
Lithify (lithified): the process of turning sediment into rock by burial, compaction, and cementation.
Metamorphic rocks: one of the three main rock types, metamorphic rocks have been recrystallized and sometimes deformed by heat, pressure, and volatile gases, but unlike igneous rocks, they have not been melted.
Natural bridge: an opening in a rock that is formed mostly by the action of running water. Streams and rivers are responsible for the erosion of the rock material and may still be flowing beneath a natural bridge.
Parting: separation of laminations along bedding planes.
Pothole arch: A less common method of arch formation, the arch starts with the formation of a pothole in the sandstone. The exposed surfaces of the sandstone layers are very irregular with small bumps and depressions covering the surface. Water collecting in the depressions may erode the rock to the underlying layer. This is a form of chemical weathering as the cement between the grains is dissolved and the sand grains are washed away.
Sedimentary rocks: one of the three main rock types, sedimentary rocks are formed by the lithification of sediments that were deposited by water, wind, glaciers, and gravity processes.
Syncline: a type of fold that is concave.
Tectonics, tectonism: deformation of the Earth's crust.
Topography: the contour or relief (difference between high and low points) of the surface of the Earth.
Weathering: all of the physical and chemical changes that happen to rock formations when they are exposed to surface or near-surface conditions.
Window: an arch whose opening in the rock is well above ground level.

9 References

Anderson, P., and the Interpretive Staff at Arches National Park, Road Guide: Arches National Park: Canyonlands Natural History Association, 31 p.

Arches National Park: http://usparks.about.com/od/usnationalparks/ig/Arches

Baldridge, W. S., 2004, Geology of the American Southwest: Cambridge Press, UK, 280 p.

Belnap, J., Soil Ecologist, USGS: Cryptobiotic Soils: Holding the Place in Place: U.S. Geological Survey: http://geochange.er.usgs.gov/sw/impacts/biology/crypto/

Blakey, R., and W. Ranney, 2008, Ancient Landscapes of the Colorado Plateau, Grand Canyon Association, Grand Canyon, AZ, 156 p.

Doelling, H. H., 1985, Map 74: Geology of Arches National Park: Utah Geological and Mineral Survey, 15 p.

Doelling, H. H., 2000, Geology of Arches National Park, Grand County, Utah, in D. A. Sprinkel, T. C. Chidsey, Jr., and P. B. Anderson, eds., Geology of Utah's Parks and Monuments: Utah Geological Association Publication 28, p. 11–36.

Doelling, H. H., 2000, Geologic Road and Trail Guides to Arches National Park, Utah: in P. B. Anderson and D. A. Sprinkel, eds., Geology of Utah's Parks and Monuments: Utah Geological Association Publication 29, 44 p.

Doelling, H. H., 2003, Geology of Arches National Park, Utah: in D. A. Sprinkel, T. C. Chidsey, Jr., and P. B. Anderson, eds., Geology of Utah's Parks and Monuments: Utah Geological Association and Bryce Canyon Natural History Association, p. 11–36.

Harris, A. G., E. Tuttle, and S. D. Tuttle, 2004, Geology of the National Parks, 6th Edition: Kendall-Hunt.

Hurlbut, C. S., Jr., 1966, Dana's Manual of Mineralogy, 17th edition: John Wiley & Sons, Inc., New York, 609 p.

Lohman, S. W. (graphics by J. R. Stacy), 1975, The Geologic Story of Arches National Park: USGS Bulleting 1393 (online at: http://www.nps.gov/history/history/online_books/geology/publications/bul/1393/sec10b.htm)

National Park Service, Arches National Park: http://www.nature.nps.gov/geology/parks/arch/

National Park Service, 2004, Arches National Park Geologic Resource Evaluation Report: Geological Resources Division, Denver, CO, 64 p.

The Natural Arch and Bridge Society: A Non-profit Society Supporting the Study, Appreciation, and Preservation of Natural Arches and Bridges: http://www.naturalarches.org

Stokes, W. L., 1986, Geology of Utah: Utah Museum of Natural History and Utah Geological and Mineral Survey, Dept. of Natural Resources, Salt Lake City, 280 p.

Wall Arch Collapse: http://geology.com/articles/wall-arch-collapse.shtml

Wilbur, J. H., 2004, The Dimensions of Landscape Arch: Removing the Uncertainty: http://www.naturalarches.org

Wikipedia: http://en.wikipedia.org/wiki/Arches_National_Park

Wikipedia: http://en.wikipedia.org/wiki/Tatooine